BUNCHES
AND
BUNCHES
OF
BUNNIES

Louise Mathews
Pictures by Jeni Bassett

SCHOLASTIC INC.
New York Toronto London Auckland Sydney

ISBN 0-590-44766-1

12 11 10 9 8 7 6   4 5/9

Printed in the U.S.A.   08

To Our Moms

Count the bunny flipping over,
By himself above the clover.
All alone but having fun,
He knows that 1 x 1 is 1.

2 bunnies proudly beam.

2 others gently dream.

Count the bunnies on a walk.
Bunny parents fondly talk,
While the baby bunnies snore.
All show 2 x 2 is 4.

3 plant seeds,

3 pull weeds,

3 spray all rows with a hose.

Count the bunnies out of doors,
Doing all their farming chores.
At the harvest they will dine,
And 3 x 3 will total 9.

4 cut a birthday cake,

4 give the gifts a shake.

4 drink ginger ale,

4 pin a rabbit's tail.

Count the bunnies staying late
To celebrate a special date.
What a mess there is to clean,
Since 4 x 4 has made 16.

5 waltz to and fro,

5 try a slow tango.

5 stand alone and sigh,

5 look spry,

and 5 hop high.

Count the bunnies at the ball,
Rabbit partners, short and tall.
Now the music comes alive,
And 5 x 5 is 25.

6 shuffle cards,

6 pull silk by yards.

6 come popping up,

6 try to guess which cup.

6 find it queer

when 6 more disappear.

Count the bunnies in the show.
Magically they come and go.
Don't they do amazing tricks?
6 x 6 is 36.

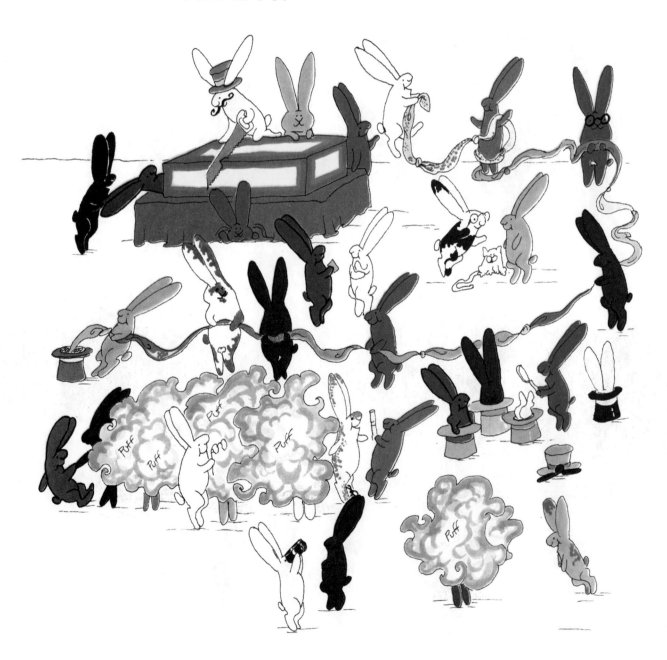

7 push a float,

7 take note.

7 hold balloons,

7 play bassoons,

7 toot,

and 7 beat,

7 lead them up the street.

Count the bunnies looking grand,
Marching to their snappy band.
Others watch, all in a line.
7 x 7 is 49.

8 swim in the ocean,

8 use suntan lotion.

8 build forts,

8 try sports.

8 ride rafts to land,

8 hide beneath the sand.

8 hear bells,

8 find shells.

Count the bunnies one by one,
Turning golden in the sun.
In the waves and on the shore,
8 x 8 is 64.

9 paste chains,

9 fly planes.

9 chew gum,

9 play dumb.

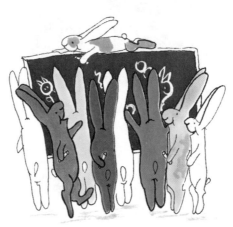

9 chalk the board,

9 pull a cord.

9 spill ink,

9 stop the sink,

9 don't know
what to think.

Count the bunnies in the school,
Breaking every single rule.
Teacher's back – they've had their fun.
9 x 9 is 81.

10 test chairs,

10 check wares.

10 wind clocks,

10 pry locks.

10 read books,

10 sneak looks.

10 make plans,

10 bang pans.

10 bid twice,

10 raise the price.

Count the bunnies who with pleasure,
At an auction, spy a treasure.
All the buyers jump to bid.
10 x 10 is 100.

11 pump,

11 jump,

11 slump,

11 thump.

11 throw a metal disk,

11 watch the athletes frisk.

11 race around the track,

11 eat a little snack.

11 handle accidents,

11 judge at all events.

11 cheer the winners here.

Count the bunnies at the meet.
Bet you cannot find a seat!
On your mark, get set to run.
11 x 11 is 121.

12 cut pies,

12 swat flies.

12 tell the news,

12 yawn and snooze.

12 pick flowers,

12 count hours.

12 toss a ball,

12 mind the small.

12 greet the host,

12 brag and boast.

12 drink punch,

12 wait for lunch.

Count the bunnies reunited,
Every relative invited.
Could that house hold any more?
12 x 12 is 144.